ALSO BY SHEROD SANTOS

ACCIDENTAL WEATHER
THE SOUTHERN REACHES

THE CITY OF WOMEN

THE CITY OF WOMEN

A SEQUENCE OF POEMS AND

PROSE BY **SHEROD SANTOS**

W · W · NORTON & COMPANY · NEW YORK · LONDON

Printed in the United States of America

First Edition

The text of this book is composed in Bembo 270 with the display set in Stencil.
Composition by ComCom, Inc.
Manufacturing by The Courier Companies, Inc.
Book design by Antonina Krass

Library of Congress Cataloging-in-Publication Data
Santos, Sherod, 1948–
The city of women / Sherod Santos.
p. cm.
I. Title.
PS3569.A57C57 1993
811'.54—dc20 92-27431
ISBN 0-393-03475-5
W. W. Norton & Company, Inc., 500 Fifth Avenue, New York, N.Y. 10110
W. W. Norton & Company Ltd., 10 Coptic Street, London WC1A 1PU
1 2 3 4 5 6 7 8 9 0

ACKNOWLEDGMENTS

Grateful acknowledgment is made to the following periodicals, in which sections of *The City of Women* originally appeared (some in slightly different versions):

Antaeus, The Best American Poetry—1991, The Kenyon Review, Shenandoah, The Southern Review, and *The Western Humanities Review.* Four of the sections ("Early morning, a woman sits up in bed . . ." and "Weeks, maybe months, have passed . . ." as *Two Poems;* "She is seated somewhere—I can't recall where . . ." as *Zorah;* and "Where to begin? My earliest memory . . ." as *Where to Begin*) first appeared in *The New Yorker.*

I would also like to thank my editor, Jill Bialosky, for her sustained and intelligent guidance with this book; the National Endowment for the Arts; and, as always, the travelers: Lynne, Ben, and Zach.

CONTENTS

THE CITY OF WOMEN

The storybook closed, the rocking stopped,
The moment of saying carrying on in a long
Duologue of silence; yet while she
Throned among her secret thoughts, the world,

Like a sunlit valley, spread from the revolving
Muslin of her skirts. A pillowing shade;
A looped-back wicker lounging chair;
And beyond the enclave of our screened-in porch,

A giant wisteria throttling the feeble
Chain-link fence. And all that endless afternoon,
The fan's rotation sweeping past in muffled,
Quickly taken breaths: A stifled sob?

Some ghost out stalking a necropolis?
Or sleep descending (though slower now
That the War is over and her loss undone)
With the sudden eloquence of speechlessness.

PART **ONE**

She is seated somewhere—I can't recall where
Exactly now, the young Algerian shopclerk
From a bookstore Mother frequented those days—
And she is seated alone, in a café, let's say,
Looking out onto a crowded square in Châteauroux,
On a market day in the early fall, a shifting
Fretwork of pushcarts, string bags, makeshift stalls,
The gutters a rubble of spoiled fruit, rinds,
Bread crusts, dung, stray dogs snuffling at
The entrails too bruised to lay out in the pans,
An acrid smell off the *pissoirs,* and the dizzying
Zigzag horseflies make in the airless crush
Of those afternoons. But all the heaped activity
Out of which I've just caught sight of her
Seems not to break upon the stillness that surrounds
Her where she's lifting her spoon to sip some
Chocolate from a steaming bowl; and though it had rained
Earlier, off and on, the sun's come out, the sun's
Reflected off the window she sits staring from,
So that her image deepens behind the pane,
Advances as if out of flame and then recedes again
Into a glassy incandescence our desolate world
Would crisscross for a moment before the next
Cloud came and shadowed her back, as in
The fade-out of a movie screen. This is a picture
I've kept for thirty years, it's always there, in a silence
Of which I've never spoken, a suspension bridge
Across that time which in some ways
Does not exist, will never exist, the story of my life
In love, the buried life I know little about,

Perhaps know nothing at all. But picking my way
Through the marketplace, that afternoon,
On my mother's arm, I've been in love for days,
For weeks, though, foolish as this may sound to you,
I didn't know that I was until, figured as she is
In the fired, unapproachable glare of that windowpane,
She seems so beautiful she frightens me.

Already, in advance, our lives owe something to those moments. An induction. A knowledge. An unlikeness between ourselves (before) and our-selves (thereafter). And a barely perceptible disharmony with all of our surroundings, as though the world had just contracted around a newly engendered set of senses. As though the world, in vast but incalculable ways, had put a DIFFERENT FACE ON THINGS.

That much we realize even then.

~

I'd met Zoë only weeks before—she had come to Paris from Amsterdam and now worked as an *au pair*—and one afternoon while we were having coffee, she asked me to tell her a story, about anything that I pleased. Because I couldn't think of anything else, I began to tell her about a little-known section of the city, three or four blocks near St. Germain—had she been there yet?—where the new, undiscovered designers go to set up shop:

Their boutiques are sparsely furnished and, because most of the designers are still quite poor, exhibit only a minimum of garments, perhaps a dozen of their trademark dresses, a smaller number of skirts and blouses, a handful of scarves and extras. So that men may shop there alone, for their wives or lovers, each boutique has on hand an assistant who doubles as a model. Most shops, comprised of a single well-lit room, have no dressing area per se, so the assistant changes behind a broad swatch of material strung, haphazard, straight out from the wall. The curtain is just a formality, for once the clothes are selected the designer's begun an excited pitch about fabrics, falls, the flares of the dress, and the curtain is normally swept aside in the course of this promotion.

When a woman shops there the custom is slightly altered since, most times, she'll try on the clothes herself. The assistant then assumes the alternate role of lady-in-waiting, helping the woman in and out of her things while the designer goes about his business with the same unabashed immodesty.

I told her one day I'd like to take her to visit those shops, to pick out clothes, to watch her dressed, and undressed in turn, in the mock-burlesque of that empty stage. When I finished my story, she said, Tell me more, about the shops and what we'd do there. But there was nothing more to tell.

Weeks later—we still weren't lovers—I went to see her at the address of a wealthy family who, while away, had hired her to look after their flat for them. As soon as I came in, she led me into an enormous bedroom with a dressing room off to one side. Two large vases were filled with flowers, and draped over a bureau against the wall there were numerous articles of clothing. She motioned for me to sit on the bed, and then, very slowly, without saying a word, she lifted off her cotton shift and began to try on all those clothes. Though not, I imagined, in the manner of a wife or lover. She dressed, instead, with the slightly studied, businesslike air of one of the shop assistants. She dressed, as it were, anonymously, and rarely even glanced my way, though her skin, I remember, gave off the faintest scent of opium, a fragrance she'd attempted to cover over with a drop of vanilla behind each ear.

~

It is a busy corner. A crush of people crossing the street, people waiting for the bus, people passing by. And a woman of forty or forty-five (she is wearing a dark-blue cashmere suit) who has lighted a cigarette and paced the few feet to a newsstand where she turns and paces back. She glances at her watch, studies the faces of the passersby, and then continues, smoking and pacing, looking in a half-dozen directions as she goes.

I have never seen that woman before. I will never see her again. Except that way. The way I've seen her ever since. Pacing. Smoking. Waiting for someone who never comes. A San Francisco street corner. A cold spring morning in May of 1972.

But how easily those figures shade, as in
A slowly developing negative, into another
Room that's charged with loss, with a woman
Kneeling on a floral rug unloading sketchbooks
From her duffel bag. And the well-dressed,
Slightly withdrawn man who moments ago
Unthinkingly said he'd prefer to live by water,
That man's standing with his back to her,
Staring out into the valley that will be their home
For the next six months: a declining sun; French skies;
A file of poplars lining the road they've just
Come down in their rented car; and a sudden,
Light gust of wind, a weak wind that's still
Warm with summer and filled with the aroma
Of lavendar, spills like ether through the tulle fog.
Though he's lost momentarily in that thinning,
Mysteriously tranquil light, the light apparently
Reminds him that it's time he thought about
Dinner now, an event he had happily anticipated
Ever since midday when he'd packed in ice
Two spotted trout, some boiled quail eggs,
And a bottle of Sancerre. And so turning around
To approach her where she's stretched out full-length,
Facedown on the floor, he asks if she'd like
Something to drink while he's setting things up
In the kitchen. But following her release from
The clinic . . . *though that couldn't have mattered less*
To her now, so fixed in her mind is the image
Of him coolly trying to persuade her that this wasn't
Really a death at all . . . following her release

From the clinic, she just can't bring herself to respond,
Which makes him pause another moment or two
Before, in good conscience, he can leave the room.

At times I'd catch L. telling me the most wildly implausible things—she is twenty-two, Catholic, attending classes at the junior college—though she never appeared the least embarrassed, nor did she seem to think that this in any way reflected on me. But her lack of compunction (if that's what it was) was compensated for by an increase in self-estrangement. At times, I felt, she really didn't know just who was speaking:

I had my first lover when I was fourteen. My parents were separated and my father was an Army Intelligence officer stationed in Saigon. Since he worked nights, my two brothers and I were left by ourselves for the better part of a year, which was fun, or at least it was fun for me. I looked older than my age, and there were no legal restrictions anyway, so I took this job as a cocktail waitress at a downtown bar called The Cave. It got its name from the entrance, which was painted to look like an enormous vagina. There was another waitress I hung around with, a white girl named Eustacia, whose father was a cook at the Embassy. After work she and I would go out to the strip, which the French had once called *Catinat* and the GIs now called *Girly,* and there we'd pick up businessmen. We were both so bored, that's what we did, night after night, for most of a summer.

And why did I think she'd made this up? When we kissed the first time, her half-open mouth. And the childlike way she squeezed my hand, as if to tell me HOW SWEET that was.

≃

I never tried to untangle her stories, to ferret out their truths and illusions, for I felt I was somehow in sympathy, even, at times, in *complicity,* with what I believed was her actual motive: A desire to sustain that deep-felt, inmost version of our lives—*that self projected onto the screen of the self*—which new lovers know empowers them (like the character of a thousand poems) and old lovers know lamentably wanes.

Where to begin? My earliest memory is dipped
In an acid of ammonia and sweat. An enameled
Box with large, weirdly illuminated numerals;
And a sweltering room where the curtains billow
Outside in on a man and woman, mid-embrace,
Who've just stopped dancing to stare at me
With a barely concealed displeasure. Leaning
Over, in a friendly way, the woman smiles,
And in a voice just slightly sweeter than she is,
Says, "Now we play some hide-and-seek." A damp
Bandana drawn from around her braided hair
(Its warm compress against my eyes) is tied
In a knot at the back of my head. A door
Clicks shut, locks—from the caustic fumes,
I can tell it's the cleaning closet I'm in—
And a faraway music washes over a second
Sound, like a *no* that's muffled repeatedly
Until it replicates a groan. I don't dare move
Or say a word. I don't dare trouble the spawn
Of light motes floating in the dark behind my lids,
Each one a face I search until the light pours
Over me, prayerlike and cool. When I come to,
The blindfold's off, the man is gone, and in yet
Another tone of voice—throaty, close up,
Edged with rum—she tells me *he* is hiding now,
"And he can see you though you can't see him."
The threat works: From that moment on, I'm aware
Of him, his eyes on me, of a presence in the world
That is sinister and unpitying. I'm far too scared
To tell anyone, and it's only years later

That my mother recalls the curtained windows
(And the Blaupunkt radio!) of our two-room house
On the coast of Bermuda; and the Lancashire maid,
Discharged early for showing up drunk,
Who had looked after me when I was three.

Following a late-night dinner and a bottle of wine—we'd returned to my apartment on the Rue St. Jacques—Zoë asked if I would do something to her.

I said yes, anything. Anything she wanted. That is what I wanted too. Whatever excited her. Whatever she most desired.

But what if it repelled you? What if, secretly, it filled you with disgust?

I told her again I wanted what she wanted. However dark it was. There was nothing she couldn't ask of me.

She said, That suits you very well, that pose. And I could see she'd decided to take it back, that she didn't believe what I had said. But already the air was charged with *it,* with what we hadn't done. It was there between us. Unspoken. Unspeakable. And memory refuses to let it go.

Zoë: " *'A yellow glove was hanging from the branch of a plane tree outside the Ministry.'* That is the sentence that formed itself as I was already more than a block away. But why, I wondered, *a* glove, *the* branch, *a* tree, *the* Ministry? In English that's not so clear. Why not just as easily: *'The yellow glove was hanging from a branch of the plane tree outside a Ministry'?* In fact, I liked that better. That little change made the sentence seem . . . what? More unlikely, I guess, more mine. As though by an act of *my* own will I had dominated over its meaning. As though for a moment there were nothing to fear if I suddenly took everything literally."

The story of a woman who will leave a man.
A man who thinks in metaphors. A woman
Who thinks the reason she'll leave has less to do
With him than her. Some flaw. Some ache for things
That end. A second nature inside of her: A land
That's seized by absence just as some lands are
By snow: the furious snows of late December,
Powerlines down, streets deserted, the bright
Abrasions the wind has scored against the loudly
Ticking windowpanes. At a complete loss
For what to do, the man's been walking through
That snow. His eyes are wet. His hair is wet.
He stands inside the door and waits, as only
The truly alone can wait, for her to say . . . say
*Any*thing. Get out. Go away. I hate you
More than ever. But while he was walking,
The woman had banked the fire again, and now
She sits, staring in, apparently unaware of him.
The new logs sputter on the open grate, the spent
Ones collapse and spill into the room an amber
Light that fills her hair, from where he stands,
With a cruel color, a shade of red: *the burn*
The arsonist dreams of just before he lights the gasoline.
But then chance has it that before the figure
Striking a match lifts flame to snowmelt pooling
At the feet of the man still standing inside the door,
The woman turns and in a heartbeat says, I'm sorry.
I'm sorry, I'm sorry, I'm sorry. And suddenly
It seems the flame's blown out, the match withdrawn,
The conflagration put off for another day,

And the day itself returning to them in a stream
Of hopelessly idle plans—the swan boat rides and
Hiking trails, lawn games, skating, a late-night
Swim—which couldn't have been more dimly met
By the three of us (who are their children), blowing
The steam off our scalded mugs at the breakfast
Table, Hôtel de la Mer, in the summer of 1963.

After the divorce, my mother moved to another state and, following a deliberate period of mourning, gradually grew more independent. She learned to use a wrench and screw driver, pump her own gas, and suddenly gardening was something for which she NO LONGER HAD THE TIME. In less than a year she stopped ending her letters: "I hope all of you are better off than I am."

While she became more cautious about certain things, like how much money she carried in her purse, she also seemed to lose her fear of going out alone. Squinting unnaturally—because she's just come out of a matinee?—she's leaning against a movie marquee for *The Diary of a Country Priest.* It's not clear who has taken the photograph, and for one childishly painful moment I imagine her stopping a stranger on the street. Someone completely unaware of the enormous effort it took for her—betrayed by the glare—to present herself at ease.

But why *had* she chosen to be pictured this way? Surely, "at this distance from the marriage," Bresson's film was intended as a joke—its austere evocation of the religious life, its useless young priest who finds that life, in any case, so spiritually unrewarding. Or was it, like a joke, meant to convey a message she appears on the surface to ridicule? The message: Don't pity me.

How humorous it seems to me, the phrase "excruciatingly untimely," which a crime-series cop—"Miami Vice"—uses to describe the death of a diplomat struck by a bus while crossing the street in Lima. Then I remember, it's the same, or similar, phrase my mother had used when she phoned the house that summer: I know this must be *excruciatingly untimely*. . . . Or some such thing. Though now, instead of funny, the cop's remark seems "ruthlessly precise."

A man and woman are sitting outside in early June.
She is barefoot; he is smoking; her hair is down.
The fireflies have found enough of a dark
Against which to strike their phosphor. A faint,
Lemon-scented candle. Castor and Pollux
Struggling to rise through the just-reneedling
Cypress tree. And out of a blue-going-blacker
Now, she asks the question all lovers ask:
What was it you first desired in me? By the time
He answers, she has leaned back heavily into
Her chair and closed her eyes, though he can't
Tell if it's in anger, impatience, amusement,
Or, as he chooses to think (for it's what he first
Desired in her) her brutal cinematic ways.

At a dinner party the woman across from me turns from her husband and, in a voice that's meant to include us all, says: I hate it how in conversations we're forever reduced to a "we." It's as though he and I were somehow united in the fundamental shame of lacking a self.

She means this remark affectionately, a sign of some struggle, both hers and his, to preserve those features of their individual lives. But why, her husband good-naturedly asks, why has she chosen such negative terms, "hate" and "reduced" and "shame" and "lack": I mean, couldn't I, with a less hidden sentimentality, say just the opposite? That I *love* it how in conversations we're forever *enlarged* to a "we," that it's as though you and I were somehow united in the fundamental joy of *creating* a self?

But rather than seeing his idle remark as anything vaguely playful, she takes it as a betrayal, a breaking of the pact all lovers seal: to accept, unquestioning, those public illusions upon which our self-esteem depends. To let him know how deep he's cut, she says (as if in confidence, to their guests), I sometimes wonder if what I love in him isn't completely made up by me. If maybe there's really *nothing* there. Nothing that's not my *own* invention.

And again the table has turned to him, and again he answers, now taking her hand, that maybe what he loves in her is completely *real,* yet wholly *unimaginable,* wholly inexhaustible by him, that there's nothing in her of him at all. . . .

And so the argument builds, as if from an instinct all its own, point and counterpoint for a quarter of an hour until, in the FULL VOICES OF A TRAGEDY, they seem to be ACTING OUT THEIR PARTS: The parts of two people who are "somehow united in the fundamental shame of lacking a self."

At the American Nostalgia Movie House in Montparnasse, Zoë buys a box of heart-shaped candies that have little messages inscribed on them. She hands me one: BAD BOY. And another: LUV YA. And a third: COAX ME. Before the curtain rises, the piped-in music: ". . . *these endless shared and tender secrets* . . ."

Don't be afraid, you have been afraid enough
For now, go back to sleep . . . a glare like river
Water sun-licked sliding down the length
Of my arm, and blood somewhere, on the backs
Of my hands, on the roof of my mouth, where
I can't quite call it into consciousness, which flares,
In any case, here and there in a bedroom dark
She at last extinguishes with a word . . . *Don't be*
Afraid, you have been afraid enough for now . . .
Now the bunched, astonished children have unpiled
The ditch in single file, now the cartoon wreckage
Of the school bus has reversed itself, and in a slow,
Ratcheted, sideways roll rightened unrumpled
On the gravel bend . . . *go back to sleep* . . . where
Sleep is the language of the newly opened book
Of trees, of the wet window the rain blows in,
Cool and syllabled and storyless as fountain spray . . .
You have been afraid enough for now . . . and now
It all begins to go, detach and whitening disappear
In a gathering flurry of family names, from an all-
But-imaginary childhood, as though the sweep
Of some familiar hand transformed the rainfall
Into snow, where Mother is calling her sisters home:
Aunt Sarah, Fanona, Harriet and Rose, Sally
Brown, Sara Crigler, Aunt Edith and Min . . . *go*
Back to sleep . . . the names all say, and so I do,
And dream that night, and for nights to come,
That I've awakened midmorning in the city
Of women, and when I awaken I'm unafraid.

≃

L. is sitting reading in a lighted cabstand. It's after 11:00. The streets are empty. She has been there three quarters of an hour, waiting for me to drive her up to visit her mother on Labor Day. Because I'm late, I pull up to the curb and brake more dramatically than necessary. Still, when I lean over and open the door for her, she pretends to be surprised to see me.

But it turns out that's not all. As we're leaving the city, she describes this drunk who'd harassed her while she was waiting. He had started from a distance, and then he came and sat on the bench beside her. He'd touched her hair. He'd called her "sugar" and "sweetness" and "lover girl." And then, for no reason, he went away. And all the while she'd pretended to ignore him, even when he'd touched her hair.

Neither of us has mentioned my being late, but whatever small pang of conscience I'd felt was enormous now, galling and unforgivable. When I begin to explain, with a wave of her hand she dismisses it all as ANOTHER OF LIFE'S NATURAL HAZARDS. At that my shame resolves itself into a sudden, moribund anger, as though the issue were no longer *the actual danger* I'd placed her in, but *the imagined slight:* That she'd forseen all along my petty burden of vengefulness and guilt.

At that age I'd been happy enough—after two
Or three weeks of backpacking down along
The Catalan coast—to have stopped for a day
In some seaside town, though it was wholly
By accident I stumbled into Sitges at carnival
Time: A spray of streamers from the quaysides
And restaurants, the sail-shaped whitecaps
Booming onto shore, soft salvos, and then,
At the sound of some imaginary gong,
Throngs of townspeople parading up the sand,
Bull-horned, goat-headed, mantled in foil,
And fueled by the racket of countless handmade
Carnival horns, a local fantasia advancing fast
On the filled, flung-open terraces and bars,
As though the riddling past had reversed itself
Through the make-believe of a crazy dream.
And yet, however transfigured those cobbled
Streets soon became, that passing illusion pouring
Up around us encircling the square was stilled
For a moment (as moments in childhood some-
Times were) when four storeys up, in a hotel
Fronting my curbside chair, a carved, life-sized,
Palanquined Virgin carried out onto the balcony
Cast a lavish incongruity on the scene below.
It wasn't that there was anything too high-born
In her being there, it was the calm, almost
Impersonal aloneness (a phantom aloneness,
A cloister that outstripped the air) which she'd
Brought to bear on the wrought-iron grillework
Overlooking the crowds. Still, one couldn't help

But admire the way she cradled her heart so lightly
In her hands, or how, in time, something within
The shy clairvoyance of her dreaming face
Suggested she too was in her mind's eye borne
On those ancient confusions overflowing the square—
If not exactly beside herself, still carried away
By that mutual human feeling she could see was true,
Municipal, and at that hour within our reaches.

PART **TWO**

It is the third country of my childhood:
A high canopy of Japanese mimosas, coconut
Palms and pepper trees, groves of banana
And twisted hibiscus; the mangrove swamps;
The shallow bay; and from the front stoop
Of our temporary barracks, a wide, irregular
Fringe of reef that circled like a blockade around
The island. The salt sea air was kindled with
The smell of tarmac drizzled onto the landing strip
Being leveled on higher ground; and all day long
Our tin house sang through its chronic, home-
Sick arguments, the diesel song of bulldozer,
Steamshovel, grader, and crane. Sheltered
In the shadow of an overhang, I'd watch her
Daily trudge across the heavily fenced-in
Garrison yard: Barefoot, filthy, lugging laundry
From her father's shack, she seemed by her
Indifference to invite me out from the walled
Daydream I lived in there. And so, one day,
I followed her. Out beyond the gateposts,
Lagging behind, down the shadeless path
To some breadfruit trees where she dropped
Her bundle, turned around, and staring straight
At me squatted on the path to take a pee hardly
Twenty yards from where I stood in the time-
Suspended clarity of the sun. A little, fissuring,
Serpentine stream (both shocking and hypnotic)
Trailed from between her dirty feet, and when
It ended she stood up fully the length of herself

And, as I remember, just disappeared.

Oh,

And a book I'd stolen the night before we'd left
The States. My grandfather's library's lewdest
Cover, a woman, naked, bent over a chair,
Her hair—in the manner of Munch's late bedside
Nudes—a wash of exhaustion and sexual despair;
A book I simply pretended to read, for its pages
Were badly mildew-stained, and its lettering
Sprawled in the Garamond of a language I only,
At best, half-understood: a foreign language,
The language of poetry, a poetry inflamed
By the starveling presence of a woman he called
Ô reine des péchés, my Queen of Sins.

These

Two figures I somehow conflated in the unphased
Tedium of those afternoons, an image of women
(An image of men's) a decade later it would take
A grave humiliation to see had rooted in me
So permanently I seemed destined to be alone.
It would bear its ineluctable messages—about
Poverty, gender, exploitation, class—but there,
On the island, beneath the orphaned blues
Of that tropical sky, the *un*real was more real
To me, a light translated through the prism
Of the body, cast upon the mind, and the mind
A fitful, bell-jarred place a world away
From that weltering green. It was in that in-
Most solitude that I'd write out nightly
Whole verses for her; and it was there each time
I'd reenact the wordless drama of our passion
Play—her squatting on the path, me caught
In her cold, contemptuous stare—that gestures

Were added, acts conceived, an obscure
And elaborate set (much more like Paris than
Paris would be) overlaid on an already vanishing
Here.

And then, as suddenly, the rain season came,
The first fever season I'd hallucinate through,
And for weeks on end the groaning tractors
Bogged in mud, rainrust bled from cracks
In the walls, and to a boy tented in the gauzy
Mesh of mosquito netting, the iron lights guarding
The workers' shacks crackled and sputtered
Throughout the night, as though the world made
Flesh had spoken at last in the broken dialect
Of a mother tongue; as though my own pitiless
Jeanne Duval had mouthed its syllables, licked them
Heatedly into my ears, in some shuttered, close,
Patchoulied flat rank with the wreckage of appetite.

≃

Imagine, for a moment, that in matters of love everything we're told is a lie. Imagine, too, how beneath those lies there's one lie central to all we're told: That love's unfolding is the process by which two people come to know each other; that the ultimate and ideal achievement of love is knowing and being known; and that knowing brings the release, at last, from our ever-present feelings of isolation in the world.

Then imagine—if only for argument's sake—how BEING IN LOVE might well depend, not on each of us *coming to know each other,* but on each of us actually struggling to guard that which knowing would give away. And imagine, moreover, how love may not be a "union" at all, but the willed preservation of that *otherness,* that sworn, unspoken Cyrano within us. Then imagine what an act of courage it is, to love and (more) to be loved: the decision to endure, *for the sake of the other,* that enormous burden of being alone. . . .

But then, we'll argue (for already we're of two minds in this): Isn't that actually the reason we love, *because* we're alone? Perhaps it is; but, if so, then isn't it also equally true that *what* we love is the other's aloneness, the unspoiled *isolato* the other is? And isn't the raw sensation of love contained somehow in the breathless surrender of *this* identity to *that* oblivion, that little Eldorado of *not-us-ness* our emotions so hungrily, so insatiably mine?

For some dark reason one could only imagine
From the large bruise above the left eyebrow,
The broken bifocals, the purse-strap snapped
And clutched like a leash in her outstretched
Hand, this Thursday during the noon hour rush
The Métro escalator delivered her up, as from
The throat of the world, to a streetside landing
Where people at first said unkind things,
Thinking a *clochard* passed out there. But then,
From a tear below the shoulder of her blouse
(A blouse which everyone later saw belonged
To someone well-to-do), a clot of blood appeared
To attract some passersby, and while one of them
Hurried to call a guard, another knelt down
To arrange her skirt and pillow her head
With a newspaper. Then it was only a matter
Of time before a small circle of people formed,
Some of them sadly turning away, others
Staring, indrawn, scared, still others studying
The face as though it were the map to a station
They'd mistakenly entered. But one among them
Has backed away ferociously, his face drawn up
In a shocked, involuntary, rancorous frown,
As if he himself had personally taken great offense,
Not *the* offense, but the sudden half-insane
Suspicion *she'd* done this to get at *him*—her skin
Taking on the gray coloration of an old wasp's
Nest, the eyes the cells abandoned yet somehow
Still disturbingly alive, as if Eurydice roused
By an ancient rage had finally risen to face the man
Who had one day driven her underground.

One night in bed, while smoking a joint, she describes their lovemaking as *hopelessly unsettled,* a sexual version of *frontier life*. She is younger than he, by twenty years, and she says what she says with a deference that (he notices now) embarrasses him.

She says: It's like we keep pushing it into the wilderness, some dark unknown, until we're comfortable there, wherever *there* is, and then we seem to set off for some new unknown. It's weird, I mean, but how can you tell if it's really out of boredom or desire?

They had turned on the television, turned off the sound, for both of them liked that flickering half-light when they made love, and now he lies watching the images cross the screen, one after the other, soundlessly. Being a man, and being aware of what that's come to mean these days, he tries to avoid taking over subjects she brings up; he tries (or so he tells himself) to allow *her* to lecture *him*.

He says, I really hadn't thought about it quite *that* way. But don't you think we're all more adventurous now than in the past? More open to sex and, certainly, to talk about sex?

But he wonders if what this is really about is how tonight they both got carried away, and afterward she had bled.

Settling against his shoulder, she too has begun to watch that soundless stream of images, that coruscating calm.

She says, But what if all this adventurousness is the opposite of what we think it is? What if it's a way of *avoiding* intimacy, and all this so-called openness is just a public stance? What if the dark's so desirable to us because *there* we don't have to face each other, there we're not at risk? I mean, what if the dark is *just* about excitement, while *real* sexual intimacy is elsewhere, in another kind of nakedness altogether?

As often happened with him, the marijuana had lent their conversation an uneasy, threatening edge—for him the downside of smoking pot—and he could feel the tide of their pleasant evening ebbing away from them. So he holds her tightly, as tightly as he can, as if trying to make two people one, and when he turns to kiss her, the light on her face (so it seems to him) is *unreal . . . dreamlike . . . otherworldly. . . .*

When the sun went down, among other things,
Only the lights from the pinball machine
Surfaced on the puddles outside the bar at 4th
And Station Hill. A suddenly truckless street—
The image of a life turned prodigal, then gypsied,
Then childishly homesick and nostalgic—
And an off-work waitress crossing this way
With what, an hour earlier, might've seemed
An utter unconcern for the traffic. And all
This time, as if time escaped him, the one offense
The drunk beside me won't forgive (*will never
Forgive,* he swears to god) is how she'd spoken
Of her *other life* without expressly referring
To him, though as she approaches he empties
His glass, as if even that might give him away.

My mother's family was Southern, affluent, aristocratic; my father's, working-class, immigrant, resettled near the hop fields in northern California. When they met in 1942, at the height of the War, they were both, so to speak, in disguise: my mother in the somber, pinstriped outfit of the volunteers at the USO; my father in the full military splendor of an Air Corps pilot on R & R.

In that sense, the haste of their wedding was a crisis; and their marriage, an ongoing version of the historical struggle between social and economic classes. Except they both subscribed to the same ideological principles: this was always a matter of REGRETTING, that of BLAMING, the other of MAKING AMENDS. In their arguments they seemed more alike, more familiar to each other, than in their moments of affection.

It was while reading a menu, taking a walk, working together in the flowerbeds, that I'd catch them sometimes speaking to each other with such exaggerated tenderness, it was like watching characters in a silent movie.

≃

On a blustery summer day, an American tourist is walking through Auteuil in the direction of the Porte de Boulogne. He is pretending to read a diner's guide while recklessly winding in and out of the people who are coming toward him. In fact, he is thinking: *How will I ever start another relationship? How will I ever begin again to listen to someone share with me those endless narratives of childhood! And how could I ever repeat my own, like the reruns of some television show?* Suddenly, into his mind there leaps the image of that orange card in the Monopoly game: "Go to Jail. Do not pass Go. Do not collect $200." That's it, he thinks, that's *just* what it's like. And at that moment he is staring into the section on "Late-night Bistros"; his jacket is unbuttoned; and his hair is tossed by a gust of wind that is whistling down the boulevard. To all appearances, he simply seems, like so many tourists, remarkably CAREFREE.

"All through lunch something's worried her,
And after, walking to the car, she begins to tell me
This dream she's had: She's a child again, she
Doesn't know *how* she knows, but she knows she is,
And she's following this trail that leads down past
Some tidewater flats and out to an estuary lined with trees,
Where she's wandering around, happy as a lark,
Not thinking of a thing, until a sound like *sheet*metal's
Shaken out over the water line, then the trees start
Trembling, the sky goes black, and all over her body
She *feels* this thing, this sudden power, like a rising
Temper she can't explain. At this point we've been
Standing in front of a store front, looking in,
And though she turns to face me then, linking arms,
And weighing her hand a little lightly in mine,
There is something she feels she *has* to say: 'You won't
Understand me when I tell you this—our marriage
Could hardly take *that*—but for the *life* of me
I've got to tell it to *someone* soon.' Though what
It was she doesn't say, and *whatever* it was,
It was already there, and there *she* is, shivering
Suddenly, covering her ears, and *saying* that thing,
Saying it now for the *world* to hear, but in
Other words, in parts of speech, the syllables
Like metal filings filling up the air, the air
Breaking down all around her, until I *too* am
Backing away, covering myself, miming the furious
Onslaught of some crazy made-up offshore storm."

One evening in spring Zoë invites me out to a magic show on the Rue Mouffetard. There's the usual assortment of sleight-of-hand—hat tricks, card tricks, mirrors and scarves—and for the final event the female assistant is placed in a long, coffin-shaped box from which her head and feet protrude. With great fanfare, the magician proceeds to drive a series of glittering swords every which way through the polished wood. When the blades come out the other side, they appear to be coated with blood. Each time (I could hear her there beside me in the dark), Zoë draws a sudden intake of breath—not as in wonder, but in a kind of reflexive physical pain.

Later that night when we made love, I had the sense she saw me now reenacting the violence of the magic show. Each time I'd enter her, that same gasp I'd heard before.

In the morning, however, it's gone:

Do this, she says, and this. I make
As if to do what she says, but already
She's there before me. Her eyes
Shut, Let me do it. Do that, she says,
And she does. And the moan, as of
A pain withdrawn, a suspiration,
A letting go:

An emptying vessel
of thought and desire.

~

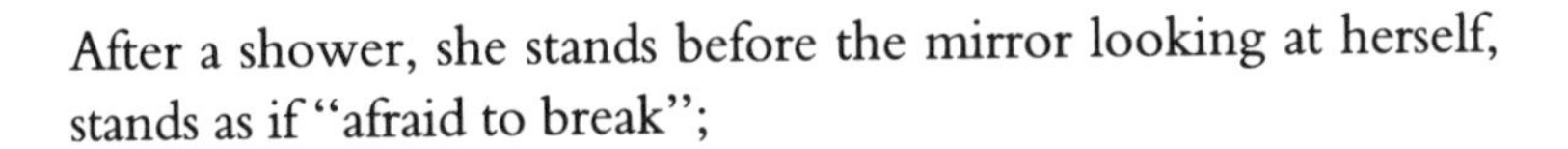

After a shower, she stands before the mirror looking at herself, stands as if "afraid to break";

---

as if sex and love dislodged each other somewhere within my body;

---

as if sex belonged, not to her at all (she's explaining this as she combs her hair), but some great primordial stream of being, passed flesh to flesh, decade to decade, age to age;

---

as if sex were something that HAPPENED to her.

Two weeks before the astronauts moved in
To the Officers Quarters down the street
On Eagan Avenue, my father in his unpressed
Khakis, just back from overseas, sat at the table
X-ing out small boxes he had crisscrossed
On a legal pad. To my child's eyes, it must've seemed
Some pilot's flight graph of the evening sky,
The way he labored over it, lifting his face
To stare straight past her aureoled in silence
At the kitchen sink. I'd later steal it from
The garbage can, and in the nightlight
Of my room study the distracted crosshatch
Splayed in galaxies remoter than I'd realize,
Until days later when the transfer came,
And with a sudden, upswept, manic air,
He loomed above her in her wingback chair,
While in furious dumb-show she let him see
She didn't know who he was or how he got there.

≃

In love—that latter-day European invention which we've never seemed to master—in love we are, depending on how one looks at it, either cursed or blessed with an impression of a world that OUTWARDLY reflects our own INWARDLY brimming souls.

And in any situation where things get taken personally (and in love almost everything does), the chances for distortion are enormous, so that our lives are played out in a state of almost perpetual hallucination, and what goes on inside of us isn't so much a *range of emotions* as a barrage of dizzying *counter-emotions,* a warring swirl of feelings we are helpless to overrule.

For example: L. doesn't cook. It's not political. It isn't something she's given up. She isn't proud of it, nor is she ashamed. It is, we agree, simply a matter of aptitude. But then one evening she's preparing a salad for the two of us, a QUIET EVENING AT HOME. She holds the tomato with one hand, the knife with the other, and slicing off a wedge she puts down the knife, puts down the tomato, and places the wedge in a salad bowl. She then takes up the knife and tomato and begins the process all over again. Only now I don't see this needless duplication, this pointlessly squandered series of motions, as "a matter of aptitude" at all. I see it in fact (for such is the weird delirium of love) as a calculated piece of domestic terror, an attempt, a willful, cold-hearted attempt, to drive me into a rage, perhaps even to the brink of madness itself.

Still, like many lovers, we preferred conflict to indifference, contention to calm. In our strict cosmology, it was crueler to say "I'm tired" than "I hate you." "I hate you," in fact, had sometimes signaled an argument's end, as though one had actually confessed one's love.

During one of our bitterest fights (the subject of which I barely recall) I left the house and didn't come back for hours. When I did return—it was well into the early morning then—I opened the closet and found her body crumpled there, PRETENDING TO BE DEAD, for while she'd waited she had fallen asleep. There ensued an infinitesimal amount of time, the time it takes a tipped glass to strike the floor, a shot bird to drop from flight, before my sudden, gutted, heartstopped gasp awakened in her a panic equal to that which she had prepared for me. . . . And we flung ourselves into each other's arms, like the sole survivors of an airplane crash.

At midnight along the Quai des Ormes the land-
Locked party barges haul against their ropes,
A procession of them lardered with four-course
Dinners, riverside wines, the pilots' shed roofs
Carpeted against the scuff of dancers stumbling
Through the dark. And fired, as in encaustic painting,
The soft wax of the acetylene lamps steams
On the cobbles and pilings, so when someone
Steps back into its glow, an image is thrown up
Trembling against the ivied arch-stones
Of the welted bridge. And it's easy to see why,
Hauled from below, their voices are still audible
To people on the banks, the sibilants skimming
Like waterbirds, bursts from the cardtables
And lighted galleys, the laughter of a girl flopped
Straddling a gangplank, now bored enough
To bother her stoned French rockstar retching
Overboard in a posture she'll find, for weeks
To come, reminds her of what she loathes in men.

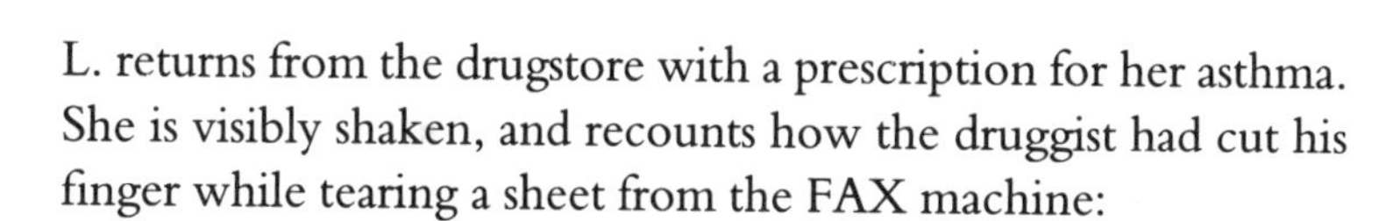

L. returns from the drugstore with a prescription for her asthma. She is visibly shaken, and recounts how the druggist had cut his finger while tearing a sheet from the FAX machine:

It was so small a cut he hadn't even noticed, until, pouring out the pills onto the counting tray, he sees a tiny smudge of blood. He then goes through this bizarre routine of throwing out the pills, wiping the tray with an alcohol swab, wrapping a bandage around this barely discernible paper cut. And all the while he has launched off into a monologue about infectious diseases, about how one can't be too careful these days, about the dangers contained in a drop of blood. And *then,* what's worse, when he's *finally* finished, he starts all over to count the pills in the *very* same tray he'd bloodied. If he hadn't said a thing, it would've been fine, I wouldn't have given it a second thought, but how can I *possibly* take these now?

She's obviously surprised by how she feels, and for the rest of the day: A benumbed look. A surreptitious faking. Like someone pretending she's "unafraid."

Angled against each other after a late-night
Party on the Upper East Side, a middle-aged
Couple in evening clothes has cut across
An abandoned lot behind a car repair: a scatter
Of shattered window glass flung in looping
Zodiacs, the pitted asphalt onyxed with oil,
A floor fluorescing in the streetlight like the scud
Of memory—a high school dance? or blanketed
Out under an open sky starred with their own just-
Unbuttoned, banked desires?—which twenty
Years of marriage has taught them was another
Life and time. And yet rising again to that same
Ungoaded loading of the heart, that same upended
Ache in the groin, they clench, kiss, hold
That way, as if testing the footing of a precipice,
Then, sure that it can stand them there, they
Climb back up toward a hungering peaked
So far beyond their reaches they can feel it
In the lungs, in the labored air, the rise and fall,
The pitch and tenor of an *O, O, O* they've
Awakened in each other as if never before
Had their mouths composed so fierce a vowel,
So pure a desperation they could cling to
Like a rock shelf in that huge, starred dark.

The curiosity with which lovers, after making love, will search each other's faces and shoulders, hands and arms . . . as though the body were somehow stranger now. Which, of course, it is. But stranger how? In what respect? And what is it, exactly, we're looking for? Isn't it actually some trace of ourselves? A bruise, perhaps? Some defect? welt? blemish? flaw? Some scar we recognize as our own?

≈

". . . the methodical, unthinking manner with which / she crosses and uncrosses her legs in public." Lines from a poem I wrote about my mother years ago. Looking back on it now I see (with a sensation of being helplessly tired) that everything else in the poem is mawkish, inaccurate, disingenuous, etc. And yet with what utter, familiar absorption I find myself rereading, while waiting for a haircut at the barber shop, one of Edith Wharton's many passages where a woman suddenly takes over a room:

"She flung her velvet opera cloak over the maid's shoulders and turned back into the drawing room, shutting the door sharply. Her bosom was rising high under its lace, and for a moment he thought she was about to cry; but she burst into a laugh instead. . . ."

Each time, a feeling of exaggerated self-importance, as though a strict confidence had just been secretly passed to me.

My neighbor is drunk. He has cut his hand. He has run out into his front yard yelling, Help me, help me, my wife is a whore. Months later (the smell of autumn leaf fires in the air) I see the two of them, arm-in-arm, walking home from the grocery store. And I am deeply moved by the resilience, by the unfathomable mystery of the human heart. And then, almost instantly (and with equal force), by its sentimentality, its cowardice, its desperate fear of being alone.

"You'd switched on the lamp. And lain down
Quietly in the middle of the light where you
Had drawn the covers off and turned your back
To me. An open window. A shade between
The reds of sunset and the ochers of a long
Twilight in mid-September swamped the out-
Lying mottling hills freaked with granite
And pampas grass. Not a breath of wind.
Not a word between us for the time it took
For you to bathe and return to bed; and if
I'd ever been happier, I don't know when
It was. Or so the mind reconstructs a guttering
Hour in summer thirty years ago, when
Something like a burnished spirit redisposed
Along the body's riddling chain of nerves escaped
Me almost out of hand, and I was poor again.
Forgive me, love, for hauling back across
These years, our wedding day (were you alive
You'd call my wanderings womanish), but I
Recall it all so clearly now: The lamplight fell
In patterns across those brightened wallpaper
Birds. Someone called out from the street below.
Then there was laughter, and many voices,
Equally light and empty, filtered on the air outside
Our room as a party spilled out from the hotel
Into the parking lot beyond. And then a fight
Broke out. I never told you this—you must've
Been asleep by then—they screamed the most
Cold-blooded things, two men, I thought, around
Whom others must've formed a ring, for there

Were shouts and threats that seemed to goad
Them on, the violence of it spurred, retributive,
Larger-than-life, as though some savage faction
Had discharged itself from the human heart
Now fired and fomented by that setting sun.
And then I feared they'd kill each other, or one
Would kill the other, and no one even try to help.
I thought of waking you. I called the front desk
And they said they'd go, but for what still
Seems the longest time, an eternity of groans
Followed by the dull thud of blows, it did not
Stop. And what I knew of love just then—
That credence I'd sworn by only hours before—
What I knew of love just then seemed like
Some foolish childhood game of make-
Believe, which everyone (everyone, that is,
Except for me) had long ago grown out of."

PART **THREE**

Squadron flags; a military band; the roped-off
Gallery of VIPs. Paired F-100s strafing
Mock-ups stationed outside the perimeter
That was cratered earlier to look like infantry
Dug in there. In the packed grandstands,
A woman and her son now sharing a soda
Find the demonstrations increasingly dull.
Just up the runway, families are climbing
In and out of the huge, down-loaded bombers.
The weather is turning. The air is filled
With the inalienable power of the coming
Winter. On the blackened tarmac,
Beneath the stands, chunks of ice the boy
Has dropped each time his mother looks away—
Sweet boredom's scattered crystal tears—
And the feeling that all this happened before

In another life, a life BEHIND the life I lived.

~

My name, L. tells me, is really Marianne. There was a boy she'd known in boarding school, the best friend of a boyfriend she'd been seeing in those days. A boy: who goes out with them from time to time, who rarely pays her any attention, and who, maddeningly, continues to get her name wrong. One day she decides she's had enough, and while her boyfriend's gone she tells him so.

The boy is crushed. Embarrassed. He says he's sorry. He says it's because she doesn't *seem* like a Marianne, he has trouble with that name. He says it just doesn't suit her, not *that* name, it's the name, that's all, *the name*. It's then she realizes it's shyness not indifference that has made him so aloof; she realizes, as well, that in his way he'd cared for her, cared deeply, perhaps, through all those weeks he'd been with them. He says he's sorry, again and again, and while he's apologizing he begins to cry, not making much noise, just tears and A LOOK RESEMBLING GRIEF.

After that he stops coming around. He was a good friend for my boyfriend, she says, but we never see him again.

The loss in her voice awakens, simultaneously, a loss in me. So Marianne is who you were? I say.

That's not how I put it to myself at the time.

But because of him you changed your name?

Oh, come on, she says (though she's faraway now), let's not go into that. That's only who I thought I was. I was seventeen, discovering myself, and sick of that name anyway.

She said nothing more about it, but from then on I believed she carried a name she'd chosen, in time, for the shy, lost boy she'd missed the chance to love. I believed as well that from then on, whenever I called her by her name, I somehow managed to return to her a part of that love she'd unwittingly destroyed.

And, in truth, I couldn't imagine calling her Marianne.

Dog days. A passing shower. Two grackles
Scattering the birdbath birds. A reshuffled
Breeze dealt off the crowns of the sycamore
Trees behind the house, where L.'s back
Out on her beachtowel reading a magazine,
Naked in the after-rain. The wet, wild canes
Of the climbing rose. Wet sunlight down her
Back and thighs (across her shoulders, hair,
Buttocks, arms), and on the freshened grass,
Unmowed for weeks, that in the shifting
August air appears to ripple like a sea surface
That's been touched from below by schools
Of brightly colored fish, rainbow fish
Passing in waves athwart the bow of the fragile
Lighter that's borne her here from the wreckage
Of that distant squall. . . . I suppose it's true,
The understory, how men can't help imagining
They're the landfall in a woman's life, but as
She turns the sun-blanked pages of her magazine,
It's as if their blankness signals me: the pathos
Of a longing, *an unlived life,* swept clean
Of all such sayable things. And I can feel
All over that closing in the chest, that old
Desire, which anywhere anytime returns to us
As the secret thrill of loving a stranger.

My mother and father went to pains to interest themselves in "hobbies," what they called "our necessary distractions." One winter—the Eibsee Hotel, Garmisch, Germany—the two of them rented complete sets of ski equipment: Decked out in fur-lined parkas (with EIBSEE stenciled across the front), oversize mittens and lace-up boots, they look mildly absurd, like shy schoolchildren who against their will have dressed as explorers for a costume ball.

On the other hand, they just couldn't be more proud, more happy for *each other* when together they push off down the hill and, in an awkward, drawn-out winter's dance, somehow manage to tip each other headfirst into a snowbank. The two of them sprawled haphazard in the snow—both laughing so hard their cheerful wreckage grows more tangled as they struggle to part—were a spectacle whose ridiculousness I felt ashamed to see. And yet I think in that moment I'd never seen them happier, as though, in falling, they'd shaken loose their proper names; as though, AT LAST, they'd managed to achieve a giddy, reflected, couple's personality.

≃

PARADOXUS. *To be one with the other,* that would seem to be fatal to the human heart (idealism); and yet, *to be wholly other,* that too seems fatal (solipsism). Thus the necessity for moving constantly between those poles; or, as in a figure 8, *through* those poles. A cycle entailing the endless destruction and reinvention of a self.

Perhaps that's why it never strikes us as the least bit odd when people speak of being in love (at one moment) as a shared experience, a reciprocity, an opening of the self; and then (at the very next moment) as the one experience to make them most acutely aware of their confinement in the world. In love we are, simultaneously, never more giving and never more needy, never more secreted and never more exposed, never more willing and never more afraid, never more lost, never more found. . . . An anxiety, indeed, a crisis, far beyond OUR WILDEST DREAMS. A crisis we find supremely desirable.

Is that not the secret passed on to us in *The Wings of Desire* . . . when the angel, who has given up his stark disembodied life to join the mortal orders . . . when the angel meets, in a nightclub bar, the trapeze artist with whom, while an angel, he had fallen in love . . . when the artist turns to address the man she has only known as an absence before . . . when at last the two of them stand there staring, FACE TO FACE . . . is that not the secret we overhear when she (to herself) *thinks to us:*

"Only with him could I be lonely."

The last to unload their bronze tonnage, the court-
Yard oaks, according to *The Georgics,* distinguish
Between true happiness and privation. A pack rat,
Black snout, little shining eyes, burrows under old
Newspapers we've left stacked beside the hallway door.
A frosted Sanskrit on the uncaulked windows.
The moon shellacked and thumbtacked to the glass.
Zoë setting aside her novel to light a cigarette—
To light it with A RUINED STARE—as though the novel
Had actually been written about her. All the uncalled-for
Et cetera at the end of autumn, and all of it scrawled
In the Old World emblems of a broken heart.

She has the clear blue eyes her doctor calls "light-sensitive," though she's someone who seems to take real pleasure in the play of light on surfaces: the sparkle of a river rinsed with clouds, polished spoons, fireworks, twilight, star charts, snow. . . . I wonder now if it's *that* my father was thinking of:

Christmas morning at my mother's parents' baronial house. Everyone dressed in Sunday clothes, each of us presented with a package while the others sit and watch. No one really wants to go next, and it's mother's turn, my father's gift whose sleight-of-hand has thoroughly fooled us all. Slowly, almost *tenderly,* I think, she removes the wrapping from what we've guessed is a bottle of champagne; only it isn't champagne at all, it's a dummy bottle from which she draws—astonished beyond the reach of words—a dress of glittering, gold lamé, like an overflow of warm champagne.

The kids are about to break into a wild applause, but it's grandfather she is turning to, as if into the weird distortions of a fun-house mirror, when she rises to say "I couldn't possibly anywhere ever wear anything even remotely like this." And that is that. Father dumbstruck. Mother humiliated. Grandfather apparently satisfied. And the children somehow oddly removed, as though we'd only served as mediums.

And yet, how often these days, struggling to recall some incident or other, I'm struck by a feeling of sifting through ash; and the adult fantasy (tinged with fear) that who we are is composed of what, perhaps only what, we can never reclaim from the rubble.

Then, as well, there are certain moments which happily project themselves forward in time: She gets in the car and opens the door for him. He gets in and reaches over to lock her door. Later, in the living room of the house where we are guests, he gets up and cuts an apple and takes it over to her; then she gets up and does the same for him. Still later, in the hall outside my bedroom, for a long time they stand there embracing. In my sleep that night their embrace appears exactly as before, though not as a PICTURE FROM THE PAST, but a DREAM OF THE FUTURE. The embrace, I remember, smells slightly of liquor. Of liquor and apples and early fall.

Early morning, a woman sits up in bed
With a cup of coffee and an ashtray in her lap,
Though she isn't smoking and the coffee
Has long since cooled. For the last two months
She and her husband have slept in separate
Rooms, and now, by habit, it's decided this room
Is "hers." Outside, the sky is overcast,
As it usually is in the mornings in the fall,
And there's a stillness on the world, which
For once she doesn't find threatening. Beyond
Her window a sparrow is furiously tearing
Away at the wildly overgrown lantana bush,
Stabbing at its inky, blue-black berries,
Some of which fall onto the window ledge
Already badly stained. Before entering
Her room—he's dressed for work and probably
In a hurry—her husband pauses and shuffles
His feet as though wiping his shoes on a mat.
At the sound this makes, she looks up at him
Undisturbed, and so manages once more
To turn a loss into the semblance of a loss.

~

For a time, Zoë and I had decided that we wouldn't see each other; and then we decide it's probably okay. The first thing we talk about is "the sex." How we both miss *that,* but not *the other,* the fighting, the recriminations, etc. We realize how impossible our relationship was, how right we were to give it up, how filled we'd been with unconscious anger at the sheer exhaustion of our physical needs.

Still, she speculates, it's too bad we can't separate the two. Ex-lovers, she's noticed, can be sexier than new ones. Sometimes, for example, she imagines that I might touch her still, touch her in ways I never did, while we are seated (as we are now) divided by a table in the Bal Musette. And nothing, she says, nothing can change those feelings, nothing we say, nothing we do to rearrange our lives. It's as though they existed outside of us, with an IMAGINATION ALL THEIR OWN.

As she continues talking, I am reminded again of how slender her wrists are, how thick and heavy her hair, which appears to draw up into itself all the unloosed ardor of her words.

≃

Another time I run into Zoë as she is coming out of a *parfumerie*. It's just past 6:00; the dark is falling. She is wearing a bright red *Chinois* scarf, and her hair is cut in a way I haven't seen before. We talk very briefly, but as we do the passersby BEHIND ME are reflected in turn in the huge store window that looms up brightly BEHIND HER. Which makes us seem—leaning together as the world hurries past around us—like two people joined in some inexorable way. Finally, in parting, I ask if she is happy—to which she laughingly says, You will have to smell me to know for sure.

She is walking away from me, well down the street, barely in sight, before I can bring myself to turn and—avoiding any CONSPICUOUS GESTURES (like one of the faceless passersby)—walk off in the other direction.

Terre vague. The room in which she sits
Is spacious and airy, and the morning flooding
Through the terrace doors lights her hair,
Lights her, in fact, in her youthful middle age.
Everything else seems rayed around her:
The bed laden with knapsacks and suitcases;
My brother and sister playing what looks like
Patience on the floor; my father sitting
In an easy chair with a newspaper on his lap.
And for whatever reason—an illusion
Of the sunlight, her immense (or too small)
Pleasure with this room—it appears her eyes
Are glinting with tears. And the fixed impression:
This is how she prefers to remember us.

L. arose one morning and began to wash the floors and windows in all the rooms of our rented house. It took her late into the afternoon, and when she finished she sat in a chair and wept out loud, because, as she said, I hadn't even tried to stop her.

Later that night she described to me how much she'd GROWN: When I was a child I would wake up in a strange bed and scream. Now I wake up time and time again and have learned to accept it passively.

In her dreams that year, *the endlessly frustrated desire to dream.*

Or: A calm, growing calmer, until it finally becomes an undesirable calm.

Or: *Joy, sorrow, boredom, grief* . . . though all she can recall is their intensity. Nothing of what they are, or why. And throughout the morning, how they *shadow her like a hangover.*

Or: A silence so fragile she feels she's actually *preserving it* by *being in the dream,* that it has been trusted to her care, and she knows that if she awakens she will shatter it "beyond repair."

Or: Certain feelings that didn't seem to work *through her;* feelings *that just kept watch.*

Weeks, maybe months, have passed, and just
Outside the kitchen door my father is standing
On the redwood deck listening to the owls,
Who call, or answer, or resign themselves
To a mounting dark in the woods beyond our road.
The more he listens, the more they call.
Seeing what his life has come to—how else
To say this now?—perhaps it's not too much
To think: Things might get a little better
In time. Time, all the while, sliding past
Like a calm sea beneath those boards. . . .
No, standing there he seems to incline
Toward something that inclines toward him:
The beginning of hope. The beginning of sorrow.
Something hunted deep within the forest
Of his affections. It's an hour he'd like
To preserve somehow, but already the dark
Has begun to lap against the lowest rungs
Of the railing. A small moving. Wind
In the trees. Salt wetness and bright stars.

≃

A woman I'd worked with off and on, in Salt Lake City, asks me to drive her to her father's funeral in southern Idaho. They hadn't spoken in several years (for reasons she was never to explain to me), and she wanted to make this last good-bye. Her mother had died when she was a child and there was no other family to speak of. The funeral took place in a little town near American Falls, where she was born, where he'd retired soon after the war. I waited for her outside in the car. She was only a few minutes—the wake was held in the backroom of a hunting lodge—and as we left, I asked her what it was like "in there."

She didn't answer immediately, and then she said, He looked like someone who'd forgotten to suffer. And suddenly I'm reminded—with a stinging shame—of another place, another time:

And again L. is saying, *This can't go on.* And she is trying to say it REASONABLY, though we both know what she really means: *What's going on, what keeps going on, what will go on no matter what we say,* is that very thing which *can't go on.*

I have no response. There seems to be nothing left to say. We've been through this, this "in-between," a hundred times before. But when she sees I refuse it at a hundred-and-one, she says to me, with a contempt that must've surprised even her, *You look like someone who's forgotten to suffer.*

Though it's only now—in the context of this other life—that I realize how cruel a remark it was. And I FEEL FOR US retrospectively.

—I can hardly pretend to forgive you for being who—or what—you are.

—First you say practically nothing, and then immediately you take it back.

—After all you've said I feel the same as I did before (her gaze turned toward the ground).

—It's as if you've stopped "listening" or "not listening" and started "unlistening"—to every single thing I say.

—I never imagined it would be like this. (Pause) *I never imagined it otherwise.*

—And then you'll say, So how do you feel, *and for hours after I'll find myself at a loss for words.*

—It's like being pregnant all over again, though I can't tell if I'm heavy with the past or big with the future.

Giving and forgiving; saying and unsaying. Thus we'd seem to arrive at periods that were wholly outside time. Ended, always, by one or the other relenting, just *saying the word,* though *the word* was random and likewise about purely external things. (This, perhaps, is the enduring part of the story: those anticipated, hence, inevitable endings. And the illusion, the towering LITERARY illusion, that we've somehow brought them on ourselves.)

If I went back and saw them as they truly were,
I'd understand, and understanding realize
(By which, I think, she means FORGIVE)
They loved each other, she loved him, and god
Knows they have tried, tried harder perhaps
Than *I* could know, though even as she
Says it she seems to doubt things happened
That way, or even "happened" at all. Still,
She keeps pushing on, pressing home this finally
Indefinable thing, and I am with her, nodding
Now, watching as her lowered profile tilts
Back deferentially, for she knows, just as I
Suspect, that this talk goes on years from now—
As if to prove there is no death? no sorrow
To lift us out of our lives?—goes on somewhere
Outside us replaying those same unsayable
Words whose syllables we are laved in,
Whose meanings keep endlessly coming to pass:

~

When they'd argue, I'd grow calm; when they wouldn't speak, I'd grow talkative: a feeling of well-being, *like taking a breath,* releasing.

⋆

The young married couple, from out of town, struggling together—because they just can't give up holding hands—to open the heavy cathedral doors.

⋆

A woman who, to keep from crying, steps off the street into a shop that sells artificial limbs.

⋆

Somewhere in the mysterious turnings of sleep, her asking her father for a glass of water.

⋆

My mother in a sundress on the first day of spring, though it's raining heavily and cold outside.

⋆

Well into the night, the light from their bedroom shining in the garden.

★

The smell of salt grass off of Half Moon Bay.

★

A bouquet of flowers (alyssum, sweet william, tea roses, fern) left behind on the concrete rail of the sailing pond, Luxembourg Gardens, Paris, 1984.

★

The SHARED EMOTION of a radio song.

★

The MOMENTARY HAPPINESS in a stranger's voice.

★

A middle-aged woman who has paused momentarily to stare at the sky—a bright contrail swept by high stratospheric winds—on an evening walk in December.